Bizarre Beast Battles

Box Jellyfish vs. Asian Giant Hornet

Gareth Stevens Publishing

By Natalie Humphrey

Please visit our website, www.garethstevens.com. For a free color catalog of all our high-quality books, call toll free 1-800-542-2595 or fax 1-877-542-2596.

Library of Congress Cataloging-in-Publication Data
Names: Humphrey, Natalie, author.
Title: Box jellyfish vs. Asian giant hornet / Natalie Humphrey.
Other titles: Box jellyfish versus Asian giant hornet
Description: New York : Gareth Stevens Publishing, [2023] | Series: Bizarre beast battles | Includes bibliographical references and index.
Identifiers: LCCN 2022024806 | ISBN 9781538283974 (library binding) | ISBN 9781538283950 (paperback) | ISBN 9781538283981 (ebook)
Subjects: LCSH: Cubomedusae–Juvenile literature. | Hornets–Juvenile literature.
Classification: LCC QL377.S4 H86 2023 | DDC 593.5/3–dc23/eng/20220712
LC record available at https://lccn.loc.gov/2022024806

First Edition

Published in 2023 by
Gareth Stevens Publishing
2544 Clinton Street
Buffalo, NY 14224

Designer: Leslie Taylor
Editor: Natalie Humphrey

Photo credits: Cover, p. 9 (hornet) akiyoko/Shutterstock.com; cover (box jelly) Danza/Shutterstock.com; series art (background texture) Apostrophe/Shutterstock.com; series art (hornet icon) CloudyStock/Shutterstock.com; series art (jellyfish icon) Nezakat Kerimova 71/Shutterstock.com; p. 5 Daleen Loest/Shutterstock.com; p. 7 David Carillet/Shutterstock.com; p. 8 Katherine Wallis/Shutterstock.com; p. 9 David Carillet/Shutterstock.com; p. 10 Dewald Kirsten/Shutterstock.com; p. 11 QueSeraSera/Shutterstock.com; p. 12 thekovtun/Shutterstock.com; p. 13 Lightspring/Shutterstock.com; p. 14 Gino Santa Maria/Shutterstock.com; p. 15 Manoj 007/Shutterstock.com; p. 16 Anita Kainrath/Shutterstock.com; p. 17 Dave Hansche/Shutterstock.com; p. 18 & 19 MrsWilkins/Shutterstock.com; p. 21 BabebiE_Gallery/Shutterstock.com; p. 21 Elya Vatel/Shutterstock.com.

Printed in the United States of America

CPSIA compliance information: Batch #CWGS23: For further information contact Gareth Stevens at 1-800-542-2595.

CONTENTS

Words in the glossary appear in **bold** type the first time they are used in the text.

A DEADLY JELLY

What is the deadliest jellyfish in the world? The box jellyfish! There are many kinds of box jellyfish. The most harmful kind is found in the warm ocean waters around Northern Australia and in the Indian and Pacific Oceans. This ocean hunter has a **bell** that looks like a harmless, box-shaped plastic bag, but don't get too close. Box jellyfish are deadly not only to the creatures they eat, but also to people.

Also called a "sea wasp," box jellyfish have 60 stinging **tentacles** with enough **venom** to stun or kill their next meal!

Box Jellyfish have 24 eyes.

THE MURDER HORNET

No one wants to get stung by a hornet. But the Asian giant hornet has a sting that really packs a punch! Also called a "murder hornet," the Asian giant hornet is native to Southeast Asia, but has been seen in North America as well. This giant pest is dangerous to other wasps, and can also be dangerous to people.

If you think you see an Asian giant hornet nest, call pest control, and stay as far away as you can!

THE ASIAN GIANT HORNET'S JAWS, OR WALLS OF THE MOUTH, ARE STRONG ENOUGH TO BREAK THROUGH SKIN.

WHO IS THE BIGGEST?

When it comes to picking the deadliest beast, size is an important factor! Box jellyfish bodies are made up of three main parts. These parts are the squishy bell on top, the arms around its mouth, and its long, stringy tentacles.

BOX JELLYFISH

TENTACLE LENGTH: up to 10 feet (3 m)

WEIGHT: up to 4.4 pounds (2 kg)

ASIAN GIANT HORNET

BODY LENGTH: UP TO 2 INCHES (5 CM)
WEIGHT: MORE THAN 1 OUNCE (28 G)

Asian giant hornets are huge bugs. In fact, this hornet is considered the biggest hornet in the world! But when it comes to size, the box jellyfish has the hornet beat. The box jellyfish is larger in both length and weight!

WHO IS THE FASTEST?

Box jellyfish actively hunt their **prey**. This means that instead of waiting, they chase down small fish for their dinner! Box jellyfish are much faster than most other jellyfish, but is that enough to beat the Asian giant hornet?

BOX JELLYFISH:
TOP SPEED: 4.6 MILES (7.4 KM) PER HOUR

ASIAN GIANT HORNET:

Top Speed: 25 miles (40 km) per hour

The Asian giant hornet might be small, but it's very fast. When they're on the hunt, these giant hornets work as a group to take down beetles, worms, and even bees! In a fight, the Asian giant hornet would have the box jellyfish beat on speed.

POWERFUL VENOM

Both the box jellyfish and the Asian giant hornet have dangerous stings, but which is deadlier? A sting from both creatures contains a special matter called venom. A box jellyfish sting can cause **paralysis** or stop a person's heart!

BOX JELLYFISH

- Can be deadly to a person in 2 minutes to 48 hours
- Number of deaths caused per year: Over 50

ASIAN GIANT HORNET

- MAINLY ONLY DEADLY IF THE PERSON IS **ALLERGIC**
- NUMBER OF DEATHS PER YEAR: BETWEEN 30 TO 50

Asian giant hornet venom is much different than the box jellyfish's. Asian giant hornet venom is made up of eight different **chemicals**. The hornet uses its venom to take down other bugs when it's hunting. Asian giant hornet venom isn't usually deadly to people, but if you're allergic to bees or wasps, stay far away!

HOW BIG IS THEIR FAMILY?

What's more dangerous than one box jellyfish? A whole group of them! Box jellyfish spend most of their time hunting alone and only group together to find a **mate**. A group of jellyfish, called a swarm or a smack, can be a few box jellyfish or even hundreds of them!

BOX JELLYFISH:
SWARM SIZE: UP TO AROUND SIX BUT MANY MORE DURING MATING SEASON

ASIAN GIANT HORNET:

COLONY SIZE: UP TO 200 WORKERS

Like many hornets, Asian giant hornets spend their whole lives in a family. This family is called a colony. The colony works and hunts together. They even **coordinate** attacks together! In a fight, the Asian hornets' teamwork would likely help them beat out the box jellyfish.

WHO FIGHTS WHO?

The box jellyfish is deadly, but that doesn't mean it isn't hunted by other animals! Box jellyfish may have strong enough venom that most animals leave them alone, but there are some animals that don't seem to care. To them, the box jellyfish makes the perfect snack!

BOX JELLYFISH:
PREDATOR: GREEN SEA TURTLE

ASIAN GIANT HORNET:

PREDATORS: NONE IN THE WILD

Unlike box jellyfish, Asian giant hornets are considered apex predators. This means that they are at the top of their food chain and they aren't usually hunted by other animals. The only threat to the Asian giant hornet is humans!

WHERE CAN THEY LIVE?

The deadliest box jellyfish can be found around Australia and Southeast Asia. While they're usually found in the ocean, during storms and mating season they can be found in fresh water as well. They like shallow, or not deep, water, but will move deeper when they need to.

AUSTRALIAN BOX JELLYFISH

COUNTRIES FOUND NEAR: INDONESIA, AUSTRALIA, AND PAPUA NEW GUINEA

ASIAN GIANT HORNET

COUNTRIES FOUND IN: NORTH KOREA, SOUTH KOREA, TAIWAN, CHINA, NEPAL, INDIA, JAPAN, AND MORE!

The Asian giant hornet can be found in a much larger **range**. They are commonly found throughout eastern Asia, especially in the mountains of Japan. They like forests with lots of trees but can survive a larger range of conditions than the box jellyfish!

WHO WOULD WIN?

Who would win this fight? Would the Asian giant hornet be able to fly fast enough to get away from the deadly sting of the box jellyfish? Or would the larger size of the box jellyfish help it to beat the smaller hornet? Would the Asian giant hornet be able to coordinate an attack against the box jellyfish? Or would the jellyfish's ability to **adapt** help it escape into any kind of water and get away?

While it might seem like there's one clear winner, in an actual matchup, who knows what would happen?

EVEN THOUGH THESE TWO ANIMALS WOULD NEVER MEET, THEIR MATCHUP IS FUN TO IMAGINE!

GLOSSARY

adapt: To change to suit conditions.

allergic: Relating to having an allergy, which is a body's sensitivity to usually harmless things in the surroundings, such as dust, pollen, or mold.

bell: The umbrella-shaped body of a jellyfish.

chemical: Matter that can be mixed with other matter to cause changes.

coordinate: To work together well and successfully.

mate: One of two animals that come together to produce babies; to come together to make babies.

paralysis: The state of being unable to move.

prey: An animal that is hunted by other animals for food.

range: The area where something lives.

tentacle: A long, thin body part that sticks out from an animal's head or mouth.

venom: Something an animal makes in its body that can harm other animals.

FOR MORE INFORMATION

BOOKS

Hansen, Grace. *Asian Giant Hornet*. Minneapolis, MN: Abdo Kids, 2021.

Phillips, Howard. *Beware the Box Jellyfish!* New York, NY: Enslow Publishing, 2023.

WEBSITES

AZ Animals
a-z-animals.com/animals/asian-giant-hornet/
Learn more about the Asian giant hornet and what makes it a tough predator.

DK Find Out
www.dkfindout.com/uk/animals-and-nature/jellyfish-corals-and-anemones/box-jellyfish/
Check out more interesting pictures of box jellyfish and learn about how their bodies work!

INDEX